Control pests

Gardening **organically**

One of the great joys of gardening is to experience the variety of life that a healthy garden contains. A garden managed using organic methods will have far more interest in it than a garden where insecticides and chemicals are used. An organic garden is a more balanced environment, where 'good' creatures such as ladybirds and beetles keep the 'bad' pests and diseases under control.

Organically grown plants also tend to be healthier and stronger than plants that rely on large doses of artificial fertiliser. In healthy soil they grow strong roots and can better withstand attack by pests and diseases. Soil can be kept in top condition by recycling garden waste to make nutritious compost. Growing the right combination of plants in the right place at the right time – by rotating where you plant your veg for example, or choosing shrubs to suit the growing conditions that your garden can offer – can deliver impressive disease-free results.

These are the basic principles of organic growing – use the natural resources you already have to create a balanced and vibrant garden. It's sustainable, cheaper than buying chemicals, easier than you think and great fun. Enjoy your organic gardening.

Every year thousands of tonnes of poisonous chemicals are dumped into the environment. These incidents are not the result of accidental or negligent pollution leaks; they are from the normal application of chemical pesticides to farms, parks and gardens and they are taking their toll on native wildlife.

However attitudes to pesticides are changing. Many more have now been banned by EU regulation and more and more gardeners are querying what else these poisons are doing to the garden in addition to killing pests.

There are many good reasons for not using chemical pesticides in the garden. The most important being the health implications for you and your family of eating home-grown veg sprayed with toxic chemicals. Furthermore pesticides may kill friend and foe indiscriminately and are quite expensive. Worse, they can pollute the wider environment and build up in the food chain to later affect birds and mammals.

All garden pests can be kept to acceptable levels with the organic method by changing gardening practice and by harnessing the helpful energies of friendly insects and other creatures.

Contents

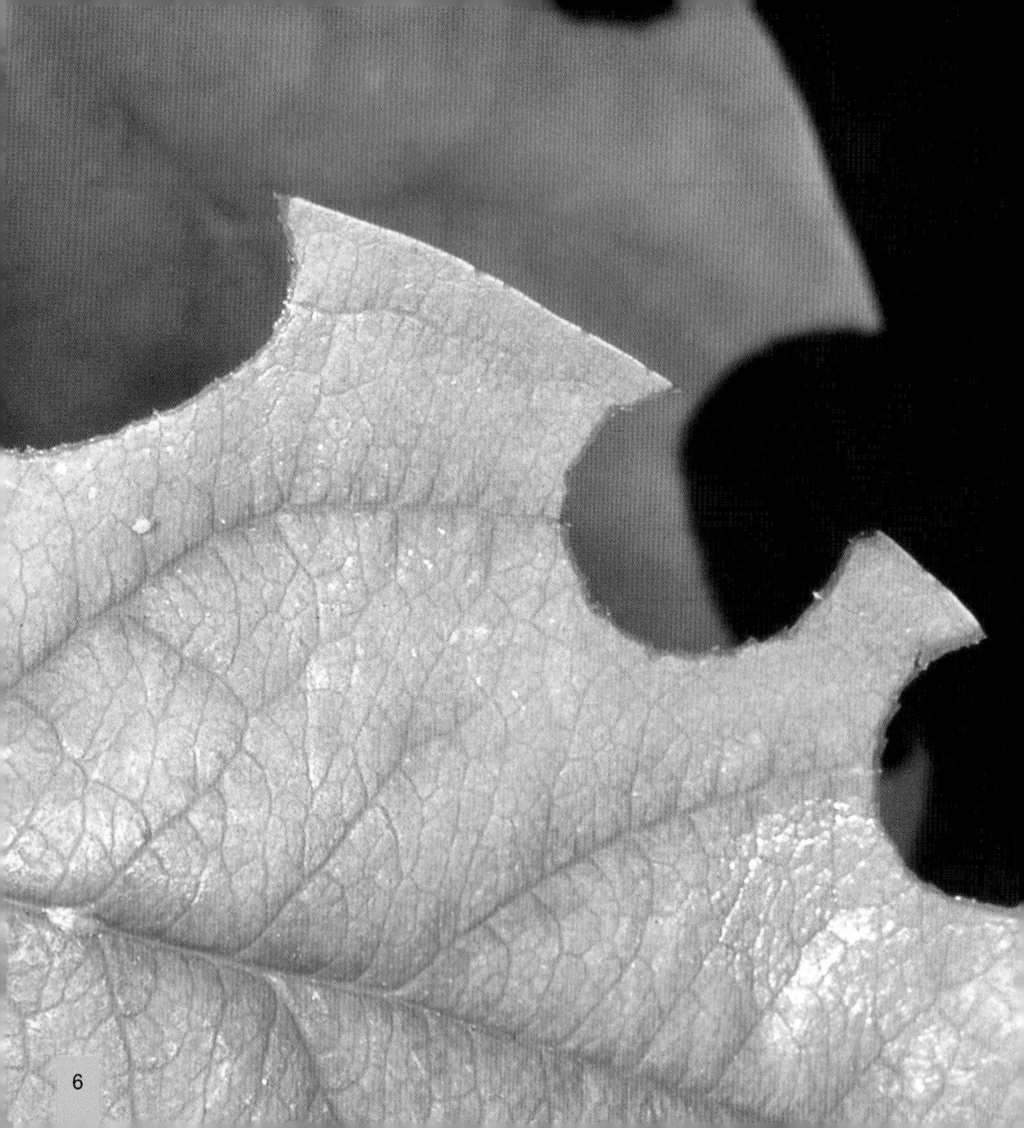

1

Why control pests naturally?

Improve your environment and **save money**

Ever since the development of DichloroDiphenylTrichlorethane (DDT), chemical pesticides have been hailed as new weapons against nuisance insects. But pesticides are designed to kill, and it was not long before their poisonous effects on non-target organisms were discovered. Now licensing regulations for these chemicals are much tougher so fewer new pesticides are coming to market and many established brands are being withdrawn amidst fears over their wider environmental safety.

The general decline of wildlife in the countryside is blamed on changing land use and modern farming practices, including, importantly, the widespread use of pesticide spraying. Spray droplets drifting on the wind can easily pollute neighbouring land, and chemicals can get into the water supply, poisoning aquatic life well away from the spray site. In gardens too, chemical poisons can spread from plot to plot despite hedges, fences and walls.

Bees and hoverflies, useful pollinators and attractive garden visitors, are particularly susceptible to insecticides, because they are highly mobile and visit a wide variety of flowers and plants. Even so-called short-lived pesticides may leave residues enough to kill these helpful insects if they come along to visit the very flowers being 'protected', or the aphid colonies being destroyed by these sprays.

Slow-release slug pellets may prevent snails and slugs eating your salad crop, but they have also been blamed for the recent dramatic decline of snail-eating birds and animals, like thrush and hedgehog.

Pesticides are expensive too. In the year 2000, for example, private UK households spent £35 million on over 4300 tonnes of chemical pesticides. The first step to controlling pests naturally is to look at the alternatives that exist.

A word of warning though, don't throw out unused pesticides with your domestic rubbish – it will just go into a landfill and the poisons will be released back into the environment. Contact your local council and they can advise you of the special pesticide disposal facilities in your area, or contact the Pesticide Action Network UK (see Resources page 61).

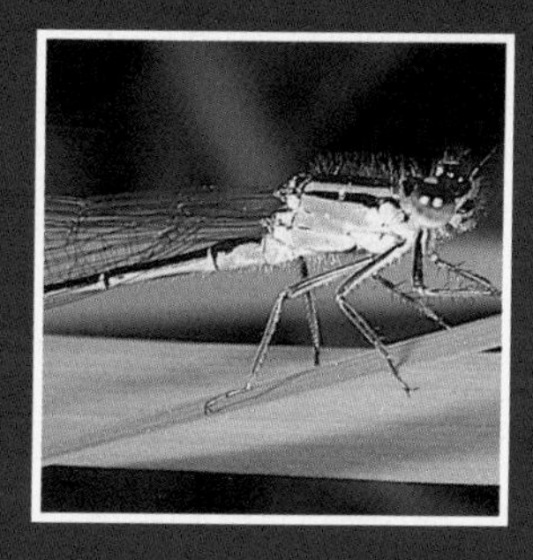

action stations

1. **Pesticides are being withdrawn from the shelves.** In 2003 EU regulations led to another 80+ common garden pesticides being withdrawn from sale; with a further 131 withdrawn from agricultural use - go organic instead.

2. **Many pesticides kill friend and foe indiscriminately.** Good guys such as bees and hoverflies are particularly susceptible due to their mobile nature - so don't use them.

3. **Pesticides and sprays are expensive.** Spend the money you save on plants. In the year 2000 alone, private UK households spent over £35 million on over 4300 tonnes of chemical controls!

4. **Chemicals pollute the wider environment and build up in the food chain.** So, before giving in to sprays, pellets and powders take a look at the natural alternatives.

5. **Now grow your own tasty and chemical-free food!**

The basic principles for controlling pests

Stop using poison

The moment you stop using pesticides, you will allow a multitude of creatures large and small back into the garden - this is the re-establishment of the natural order. These may be the harmless but brightly attractive and obvious ones, such as butterflies; or the actively beneficial ones, like spiders and hoverflies. There will be others that you are not sure about, but by observing them closely you can soon decide whether they need attention. In fact most insects in gardens are secretive and completely harmless, but all of them have fascinating life histories.

There will be an increase in the number of fascinating creatures that visit or live in your garden once you stop using pesticides.

Give your plants **the best start** in life

The stronger the plant the better it will resist attack by pests.

The key to vigorous and attractive plants is to give them the best chance of establishing and maintaining themselves, and this starts with the soil. Use garden compost and leafmould in preference to artificial fertilisers; feed the soil not the plant. Use lime to counter acidic soils, and horticultural sand and grit to break up heavy clay soils. Use a pH testing kit available from your garden centre (with instructions) if you're not sure whether your soil is alkaline or acidic.

Dig over the soil if necessary, to increase aeration ad improve drainage, removing dead roots, pieces of buried wood or large stones that might be giving shelter to slugs. Regular digging is rarely necessary, particularly on light soils. It can actually damage the soil structure if done too often. You shouldn't need to water the soil unless you encounter a very long hot and dry period, but avoid water-logging the soil if you do water.

Finally, put a layer of mulch on top of the soil - mulch is nothing more complicated than organic matter such as bark chips or rotted leaves. Mulches discourage weed growth, prevent moisture loss and act as a home to beneficial centipedes and insects.

Natural fertilisers and tonics

There are several organically certified products available that can help build up natural resistance in plants. These include:

Seaweed extract which contains a number of trace elements and natural plant growth stimulants – these combine to promote healthy plants which then become more proficient in their uptake of nutrients and consequently less susceptible to pests and diseases.

Seaweed and iron for specific iron deficiencies encountered on chalky soils.

Liquid comfrey – a good all purpose vegetable feed that works particularly well with peppers and tomatoes.

In addition there are a number of general organic liquid feeds and fertilisers aimed at specific crops such as tomatoes or formulated for general use on plants, flowers and vegetables.

But, be careful to ensure that any product you purchase is certified for use in organic gardening – ask your supplier if you are in any doubt.

Choose well

No two gardens are the same and it is difficult if not impossible, to precisely recreate a garden seen elsewhere. You must choose the plants and cultivars (varieties) that are suitable to your own particular situation, taking into account the type of soil, slope of the ground, direction of the sun, annual rainfall, annual temperature range, wind direction and strength. By applying these guidelines, you will be following a key organic rule: prevention is better than cure.

If you have previously had a problem with a particular pest try other more resistant plant varieties that may be on offer. Almost every type of plant has a range of different types available, including drought and pest-resistant varieties. Even hostas, usually notoriously prone to attack by slugs and snails, are available in tough slug-resistant forms.

Buy seeds and plants from reputable suppliers - you can even buy certified organic seed. Make sure seeds are not past their 'sow by' dates. Avoid knocked-down prices and end-of-season sales, they may be left-overs that have missed the ideal planting time and will start at a disadvantage the moment they go into the ground. Carefully follow the correct advice to overwinter or store seeds, bulbs and corms (roots).

Plant well

Make sure you plant well by giving each plant precisely what it needs. Carefully follow the directions given for germinating and sowing seeds. Don't let them grow too close, but thin them at the right time to allow the most vigorous ones the best chance of getting established before any pest has a chance to attack.

When putting in bedding or bigger plants, choose the situation that will suit them best, taking into account their preferences for sun or shade, moist or dry soil, and their resistance to wind, cold, drought or pest attack.

Remove weeds to avoid unwelcome competition for nutrients, and keep an eye on plants to spot any early signs of attack, disease or other problems.

Make your garden **wildlife friendly**

Lots of creatures eat pests – so attract them. See pages 34-37
for more information about insect predators, and pages 50-60 about how to maintain a pest-free garden and encourage allies.

A good start is important for plants to establish themselves. Start by choosing plants most suitable to your soil and the situation in which they will be planted.

Crop **rotation**

For the allotment or large vegetable patch, crop rotation is a vital technique for combating soil-living pests and diseases. By planting different crops in different places each year a build up of some plant-specific pests can be avoided; this is a fundamental principle of organic growing, following the 'prevention is better than cure' rule . For best effect, a standard rotation of three or four years can be used, dividing the land area into three or four and planting a different vegetable type or family in each zone each year. This leaves two or three years of alternative growth and soil treatment before the same crop gets back to the same plot again. See *Green Essentials Grow Vegetables* for more information on growing methods.

Because organic plants are not artificially protected by pesticides, they tend to produce more phytonutrients. These naturally protect the plant against pests and diseases, and when eaten provide us with health-giving antioxidants.

Make sure you plant well: Follow directions for germinating, sowing and planting out. And give plants a kick-start by providing them with well prepared soil, a suitable location and hands-on aftercare.

For the allotment or large vegetable patch, crop rotation is a vital technique in organic gardening. By planting different crops in different places each year a build up of some plant-specific pests can be avoided.

action stations

1. **Stop using chemicals.** Allow insects to return to your garden and for a natural balance to prevail. Most insects are harmless and the predators will help to control any pests.

2. **The key to giving your plants a good start** begins with the soil. Use home-made garden compost and commercial brands of certified organic composts. Test your soil with a pH kit (available from your garden centre); counter acid soils with lime and use organic matter and sand or grit to help break up heavy soils.

3. **Choose well and plant well.** Select plants that are suitable to your own situation, taking into account soil type, slope, sunlight, and cover from the elements that may affect rainfall and wind. Following directions for germinating, sowing seeds and planting out. Weed if required and keep an eye on them to spot any early signs of problems.

3

Barriers, deterrents, traps and controls

The most simple barriers can prove very effective. Above, a yoghurt carton is utilised as a mini-cloche.

Barriers

The simplest way of preventing a pest attack is to prevent the pest arriving at the plant in the first place. Barriers are easy to make:

- Use **wire netting** and **chicken wire** to prevent birds attacking soft fruits and seeds and to protect against rabbits, squirrels and cats.
- A string of **CDs** is an easy home-made bird-scarer.
- Fine **nylon mesh** will prevent cabbage white butterflies egg-laying on brassicas.
- **Fleece** over carrots and matting around cabbages will discourage cabbage root-fly.
- A simple barrier defence can be made from a **plastic drinks bottle**, cut in half and used as a mini-cloche on individual seedlings or cuttings. It also prevents excess water loss and if the stalk of the plant grows through the mouth of the half bottle, then the plastic sleeve can simply be cut away.

Netting will help deter attack from insects, preventing butterflies such as the cabbage white from laying their eggs on your plants.

- **Slugs** need a moist surface across which to glide their wet slime trails, so slime-resistant barriers will deter these, the most feared and hated of garden pests. **Copper tape** rings around terracotta pots, **ash** from wood fires and certain proprietary 'drying' slug barrier **granules** are all disliked. In addition, crushed garlic mixed with water may prove beneficial. For more information on slug and snail control see the *Green Essentials* book *Banish Slugs* for definitive guidance on deterring these most persistent of pests.
- A band of grease about 7 cm wide painted around fruit trees will prevent females of the winter moth climbing up the trunk. They are wingless and get stuck in the grease as they try to climb. But the grease you use has to be fruit tree grease; and trees younger than 3-4 years old should have a greased paper band instead as the grease can harm young trees.

Traps

Luring pests to a trap and then disposing of them is the perfect way of protecting your plants without harming beneficial creatures and without damaging the environment.

Slugs and snails. A shallow bowl, such as a margarine tub, filled with beer, fruit juice or milk attracts them and drowns them. Set the trap proud of soil level by an inch or so to stop the beetles that eat slugs from falling in.

Ready-made slug traps are available in garden centres. Alternatively leave out planks of wood, and examine the undersides each morning. Dispose of what you find by dropping them into a bucket of hot salty water. If you feel squeamish about killing them, take them well away to a local open space such as a nature reserve or railway embankment. Don't just chuck them over the fence, they will crawl back the following night!

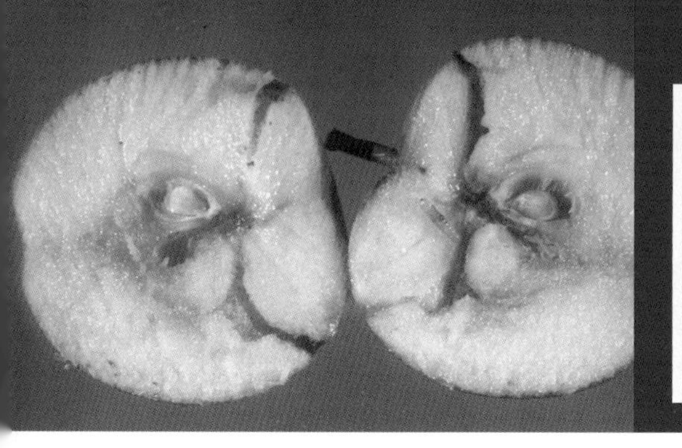

Codling moth damage to fruit (left) may be reduced by using pheromone traps; slugs and snails can be collected in simple beer traps (opposite, top); while yellow sticky traps will capture green and whitefly in greenhouses (opposite bottom).

Earwigs. Are generally useful predators, but they can damage prize blooms.

Woodlice. Sometimes a very minor pest in greenhouses where they chew new seedlings. Leave morsels of food (potato, cheese, strawberries, orange peel) under a plank or tile and examine it each morning.

Codling moth. Caterpillars burrow into apples and pears ruining the fruit. Pheromone (scent) traps hung in the branches in spring attract males, which then get glued to the sticky boards and are therefore unable to mate with any females.

Whitefly. In greenhouses, use yellow sticky traps to catch the flying adults.

Controls and treatments

Hand-picking

The best form of pest control is targeted hand-picking. This is ideal for large and obvious leaf-feeders like slugworms (a type of legless sawfly larva) and other sawfly larvae, cabbage white butterfly and many different moth caterpillars, lily and asparagus beetles and their grubs. As well as keeping an eye out for the beasts themselves, look out for the early warning signs of nibbled leaves and stalks.

The gooseberry sawfly larvae, a common pest, starts life hidden in the centre of the bush, so often goes unnoticed until the damage is done. Growing the plants as a cordon or fan, trained against a fence, makes the pest easier to spot, and remove.

Look out for the early warning signs of nibbled leaves and stalks.

Vine weevils come out at night, so search for them by torchlight on walls, paths, pots, planters and stones.

On damp evenings, or after rain, hand-pick snails and slugs. If you are squeamish about touching them, use rubber gloves or skewer them with a hat-pin or nail taped onto a stick.

The hose

For infestations of whitefly, aphids or mealybug, a squirt of water from a hand-held mister or hose-pipe will dislodge them and they will perish.

The last resort

If you feel you have to resort to sprays, powders, pellets and other chemicals, choose ones which are environmentally less damaging than the general pesticides which are available in every DIY store.

If you do resort to help from sprays, powder and pellets, there are a few products available which are environmentally less damaging. Read the label and ask for advice.

A word of **warning**

Under 1986 MAFF legislation it is illegal to concoct any home-made spray to control problems in the garden. This means that even a mix of washing-up liquid and water, (a traditional preparation commonly prepared against aphids) is now illegal.

Other old remedies such as sprays made from rhubarb leaves, elder flowers and so on, are also illegal. This is because no 'home brews' have ever been tested in the environment to see what effect they may have - although seemingly harmless, we don't really know.

Although it is illegal to concoct home-made sprays, natural help is at hand. Allies such as the ground beetle (above left) and ladybird (above right) will help to control slugs and aphids. In addition, you can buy-in microscopic help in the form of nematodes and bacterium of which there are a number of varieties for controlling pests.

Biological control

As well as encouarging natural pest eating creatures to visit your garden, you can also buy various beneficial creatures to augment the natural supplies. Known as 'biological controls' they are available mail-order, and from some garden centres.

Nematodes are microscopic worms that attack and kill pests. Different nematode species are available to attack slugs and snails, leatherjackets (cranefly maggots) and vine weevil grubs. The preparation, usually in a moist clay substance, is added to water and applied to the soil with a watering can. The nematodes will not harm larger predators such as birds and hedgehogs when they feed on the pests.

In the greenhouse: Whitefly, red spider mites, mealy bugs and aphids can all be controlled undercover by introducing the appropriate biological control agent. The creatures you introduce are minute, so you are unlikely to notice their presence – and they won't harm pets. For best results, release the biological control as soon as you have seen the first sign of the particular pest – but not before. Check that your greenhouse will provide the right temperatures for the creatures to work.

action stations

1. **The most simple way of preventing attack is to stop the pest arriving at the plant.**
2. **Plastic bottles and cartons** may be used as a mini-cloche on individual seedlings and cuttings. **Fleece and matting** keeps carrot root fly at bay.
3. **Wire netting** will help prevent attack from birds, mice, rabbits, squirrels and cats; while **fine mesh netting** will protect against insects. **Slugs and snails** can be captured in simple beer traps or deterred from pots with copper tape and barrier granules.
4. **Pheromone (scent)** traps can be used to counter certain pests such as the codling moth which attacks the fruit on apple and pear trees.
5. **Hand-picking** is ideal for targeting and controlling large and obvious leaf-feeders. Tapping pests onto newspaper and a squirt of water from a hose is more suitable for smaller pests.

4

Pests – the good the bad and the ugly

What is a pest?

Just as a 'weed' is a plant growing in the wrong place so, the true test of a pest is whether it is actually going to do any real damage to your plants. If you see an insect or other creepy-crawly on a leaf don't immediately jump to the conclusion that it is doing damage.

Insects don't just visit plants to eat them, suck their sap, or lay their eggs on them. They sit on leaves to sun themselves, to watch for passing prey, to suck up spilled pollen and to guard a territory. Before you act, get to know which is which.

Most small creatures in the garden are more or less harmless such as the shield bug (right) which feeds mainly on wild plants.

The true test of a pest is whether it is a creature that is going to do any real damage to your plants.

The good

Ladybirds. Many different spotted species, red, yellow, black and pink. Larvae grey, black or yellow, speckled with tufts. Adults and larvae are voracious aphid feeders.

Hoverflies. About 100 types, mainly the yellow and black wasp look-alikes, have aphid-eating larvae. Adults visit flowers so are important pollinators too.

Centipedes. Long-legged species have 15/30 pairs of legs; short-legged species (up to 70 pairs of legs) are mainly subterranean. One pair of legs per body segment where millipedes have two pairs per segment. Front pair of legs sharp, scimitar-like, reach round to front of head to act as jaws.

Social wasps (yellowjackets). Can be a nuisance on fallen fruit and round picnics in autumn but are usually important predators catching aphids, caterpillars and flies for their brood. Large nests only used once. Get professional help if you need an active loft nest removed.

Solitary wasps. Make small burrow nests in the soil, dead wood or hollow bramble stems. Catch and kill many types of insects to stock the nests.

Ground beetles. Many black or shiny species hunt and kill small soil-living creatures including fly maggots, slugs and insect eggs. This species has specialised narrow head and thorax to get into snail shells to attack the soft mollusc body.

Spiders. Prominent in their orb webs in summer and autumn, garden spiders catch flies, aphids, bugs and other flying creatures. Many other spiders, like the zebra spider don't spin webs but hunt on the ground or walls for their prey.

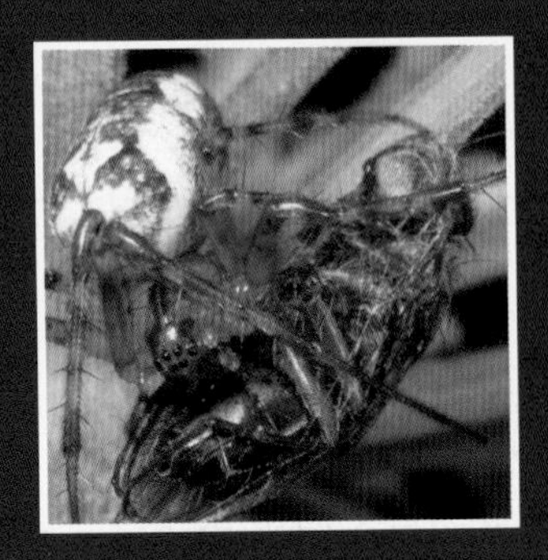

Lacewing larvae. Long curved jaws catch aphids and other plant lice. Some species disguise themselves by covering their bodies with debris.

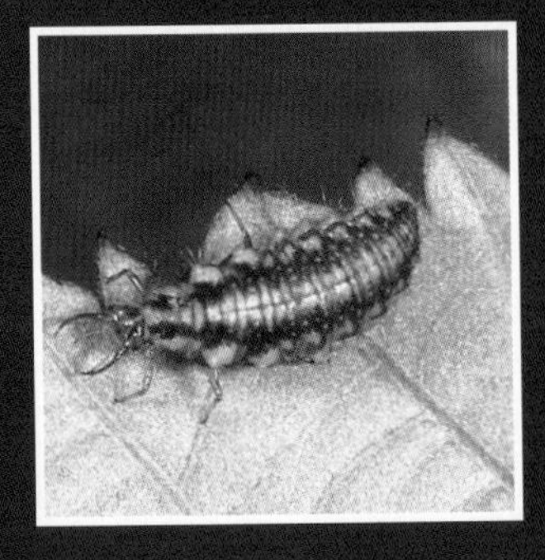

Tachinid flies. Very bristly flies that lay their eggs in caterpillars and other insect nymphs. Hatching larvae devour the host from the inside.

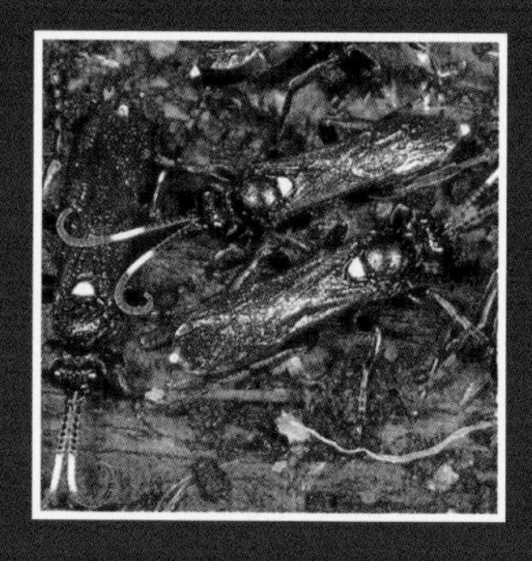

Ichneumon wasps. Usually wasp-like, yellow, red or black, sometimes with long thin tail. They lay their eggs inside caterpillars and the hatching larvae eat the host from inside. Minute species attack aphids, leaving empty hollow husks.

Devil's coach-horse. Large black beetle found under logs and stones. Has large jaws to catch and kill other invertebrates. Raises tail threateningly, but is harmless.

Flower bug. Small, often found on flowers. Attacks aphids with its pointed mouthparts.

The bad – major pests

Cabbage white caterpillars. Large white, distinctive yellow caterpillars speckled with black. Feed gregariously (as a group) creating more damage; also soiling with their frass (droppings). Distasteful to birds.

Vine weevil. Larvae feed at roots, especially in pots. Adults flightless, but good walkers. Parthenogenic, all females lay eggs without needing to mate, so single specimen can create major infestation. Several new, species recently introduced from central Europe in plant imports.

Wireworms. Larvae of slim brown click beetles, so called for the audible click they make when they move away. Subterranean, attack root vegetables so difficult to monitor.

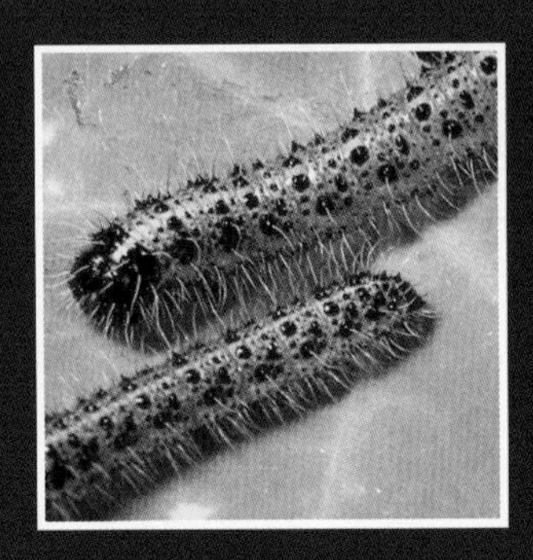

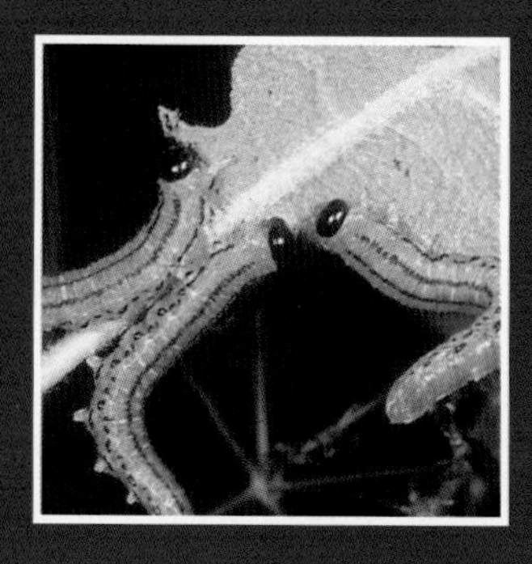

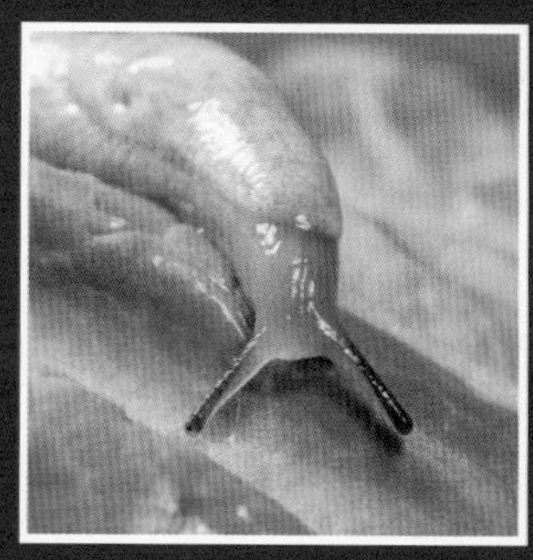

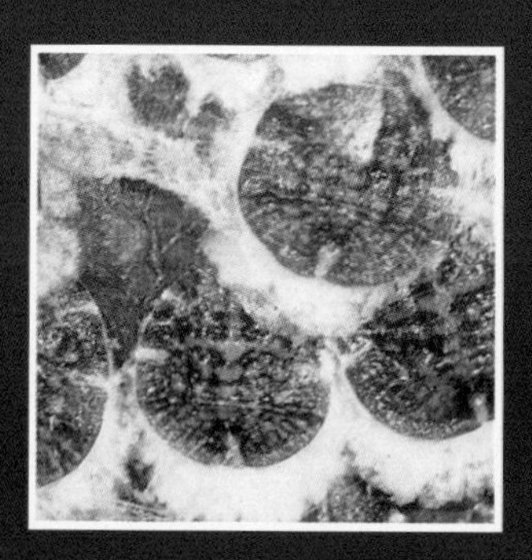

Sawfly larvae. Many-legged caterpillar-like or legless (slugworm) types. Often feed gregariously. Major attacks on roses, pear, solomon's seal. Look for herringbone egg-laying scars on stems.

Slugs and snails. Love soft fleshy plants like hostas, begonias and seedlings. Mainly nocturnal. Slugs are mainly subterranean, with no shell to hinder them, they can squeeze into the smallest of spaces. Check log-piles and flower-pot clusters for sheltering crowds. See the *Green Essentials* title *Banish Slugs* for further guidance.

Scales. These are the females of a group of sap-sucking insects related to aphids; males are secretive tiny winged things. The scale is a shield-like protection for the egg masses produced later.

Leatherjackets. Cylindrical leathery larvae of craneflies. Feed on grass roots and major infestations can damage lawns.

Aphids (greenfly). Many different species with complex lifestyles, often involving a completely different host plant during autumn and winter. Very fast reproduction means an infestation can appear suddenly. Ants are often attending the aphids, collecting the sugary honeydew produced and protecting the aphids from enemies in return.

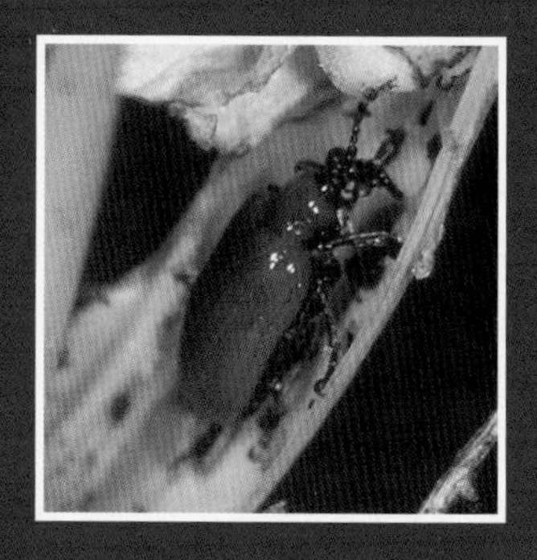

Lily beetle. Larvae shred lily leaves and flowers and spoil plants with their copious slimy excrement. Once confined to small area in Surrey, it has now spread across all of England.

Millipedes. Eat mainly plant material that is already dead, but have been known to nibble seedlings. Not to be confused with helpful centipedes (see page 34).

Don't confuse helpful centipedes (page 34) which prey on other bugs with millipedes (middle and bottom) that eat leaves and plants.

The ugly – spotting signs of damage

Leafhoppers. Small white dots on leaves are air-filled spaces where the hoppers have sucked out the contents of individual cells with their pointed mouthparts. Large infestations can cause leaf mottling, but they are really only minor pests.

Spittle bug. Small blobs of white froth on herbs and shrubs protect the pale green nymph from heat, water loss and predators as it sucks out plant sap. Mainly on native plants, and rarely enough to make it a major pest.

Leaf beetles. Nibbling starts with grazing off a layer of leaf, leaving translucent windows. Larger grubs chew the leaf margins. Grey beetles and their black larvae attack viburnum, shiny green beetles on mint. Moderate to minor pests.

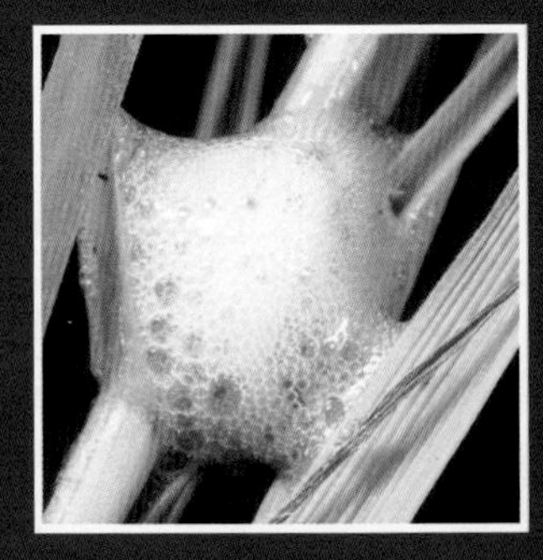

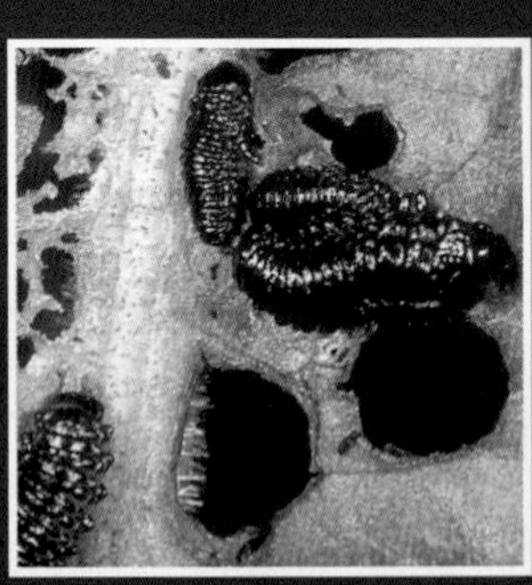

Slugs and snails. Ragged holes in the middle rather than at the edge of the leaves are made by the molluscs' rasping file-like mouthparts. The leaves are usually covered in the remains of their slime trails too.

Vine weevils. Leaf edges nibbled like stamp perforations. Particularly on shrubs like rhododendron and hydrangea. Grubs feed at roots, especially pot plants, causing sudden and dramatic plant failure.

Not so ugly: Leaf-cutter bees. Semi-circular cuts made in leaf margins, especially roses, are used to line a tunnel nest in wall or soil. Bees occur in low numbers, and are pollinators, so not really a pest.

Harmless

– don't worry about these

Most small creatures in the garden are more or less harmless. Here is a selection of common insects seen in gardens, on leaves and flowers, but which can be ignored.

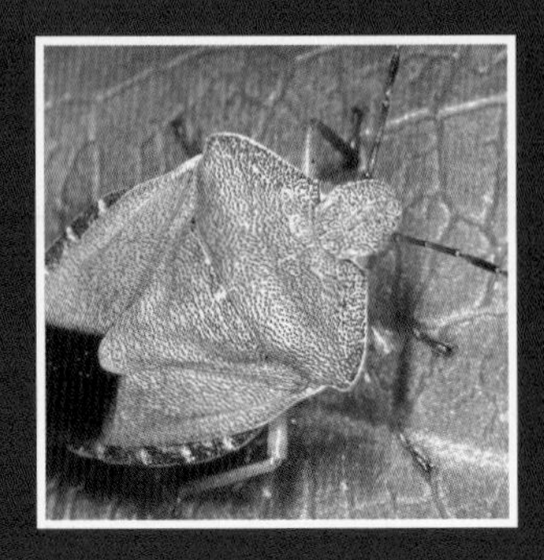

Green shield bug. One of several similar shield-shaped bugs that feed mostly on wild plants. Turns bronze-brown in winter, bright green in summer. Flies with a loud buzz.

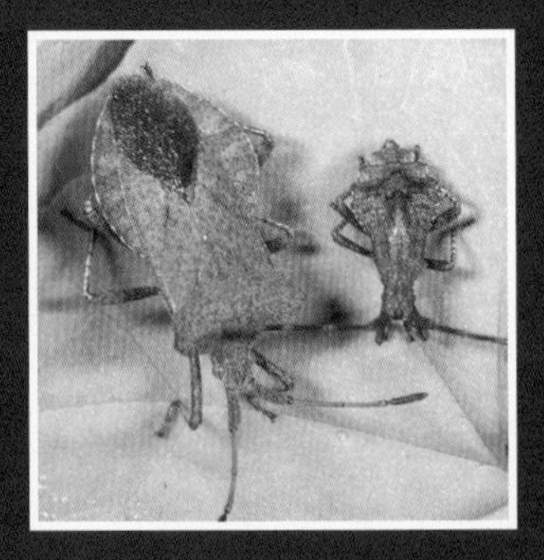

Brown leafbug. Sits on leaves sunning itself. Feeds on wild dock plants. The young (nymph) bug (right) has curled tail and looks like part of dead leaf.

Juniper bug. Used to be scarce on chalk downs, but now feeds on cypress trees (but not leyland) planted in parks and gardens. These are such vigorous trees that no damage is done.

Grasshopper. Although related to locusts, there are never grasshopper plagues. Lives in long grass. 'Sings' by rubbing back legs against wings.

Bush-cricket. Like a grasshopper, but with very long antennae. Usually in long grass or on trees and bushes. 'Sings' by rubbing wings together.

Lime hawkmoth. Like most moths this is an 'outdoor' species, feeding on wild plants and flying at night. Only a few small brown 'indoor' species eat clothes.

> *If you see an insect or other creepy-crawly on a leaf don't immediately jump to the conclusion that it is doing damage. The insects featured on these pages are all harmless.*

Yellow dungfly. Maggots feed in animal droppings: horse and cow in the countryside, but also in dog and cat dung in gardens. Does not come indoors, is not attracted to food and not implicated in the spread of diseases. Instead does a useful job recycling nutrients.

Stag beetle. Despite its terrifying size and huge jaws (only in the males) it does not bite (unless you put a finger in the wrong place). Breeds in rotten tree trunks and stumps. Flies with a very loud rattling buzz.

Wasp beetle. A wasp mimic, but perfectly harmless. Grubs feed in dead logs and twigs. Flies readily and visits many garden flowers.

Harmless: Although related to the locust, the grasshopper is not a pest. If you have a wildlife area in your garden or, are not too harsh on your lawn you may be lucky enough to find them.

action stations

1. **Get to know the good from the bad and the ugly from the harmless.** Don't assume that all creatures will cause damage.

2. **Good guys include** ladybirds, hoverflies, centipede, social wasp, devils coach-horse, dragon flies, frogs and toads.

3. **Bad guys include** cabbage white caterpillars, aphids, slugs, vine weevils and sawfly larvae.

4. **Pest indicators include** small white dots on leaves (leafhoppers), blobs of white froth (spittlebug), ragged holes in the middle of leaves (slugs), leaf edges nibbled like stamp perforations (vine weevil).

5

Maintaining your pest-free garden

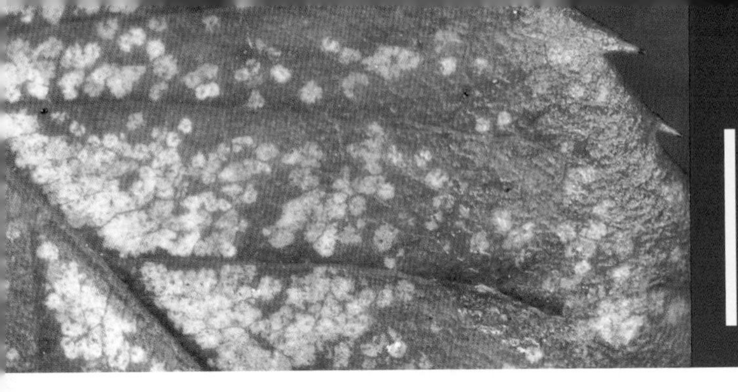

Vigilance is the watch-word of the pest free garden. Check the garden regularly for signs of attack.

Vigilance and hygiene

Vigilance is the watch-word of the pest free garden. Check the garden every day to look for chewed leaves, blotched foliage or wilting stems. Carry a small container with you so that any snails, slugs or other pests can be collected up and dispatched.

- Start checking early in the season so that you find the first hatch of tiny sawfly larvae before they start defoliating the plant.
- Plant judiciously, avoiding plants known for their vulnerability to insect attacks. Interplant a wide variety of species and avoid monocultures; that way even if the occasional pest comes out on top you have not lost an entire border or a whole corner of the vegetable patch.
- Take specific protective action to help newly established plants and sprouting seedlings, using barrier methods.Use crop rotation to avoid soil-borne pests and diseases taking hold.

- General hygiene in the garden will help prevent the spread of diseases. Don't let shrubs grow too dense, prune them to allow better air circulation and this will prevent fungal diseases like powdery mildews.
- Wash boots and tools to prevent the spread of disease and scrub pots and trays to remove any sheltering pests. Give the greenhouse a good early spring clean to get rid of any sheltering woodlice.
- Act cautiously before accepting notoriously disease-prone plants like onions or cabbages from helpful neighbours. You might get vine weevil larvae on the roots of ornamentals.

Finally, if you really have too much trouble with a particular pest on a particular plant, consider giving up gracefully and growing something else, rather than resorting to the chemical pesticides which upset the natural balance in the first place.

If you really have too much trouble with a particular pest on a particular plant, consider giving up gracefully and growing something else.

Encouraging **allies**

The most important way of encouraging your allies is by not mistakenly persecuting them. A simple identification guide will allow you to name many of the common insects and other invertebrates found in gardens. You will find that most of them are completely harmless, and that many are actively helpful predators and parasites attacking the true pests.

If you find a bug sitting on a prize bloom, don't immediately imagine that it is doing damage. Unless you recognise the beast as a known pest, it is probably just sunning itself quite innocently.

Your allies will be further encouraged if you provide some habitat in which they can live and breed (venturing further into your garden to feed on pests). You don't need to turn your whole garden over to wildlife to achieve a natural balance – a small area will do – along with a sympathetic approach to planting and management and an acceptance that (like nature) not everything always needs to be neat and tidy.

A little planning, some nesting boxes and the creation of some suitable habitat will help to encourage allies such as hoverflies, ladybirds, beetles, birds, dragonflies, frogs, toads and hedgehogs.

Help from wildlife

Birds are major insect predators, especially blue-tit, great-tit, thrush, robin, wren and blackbird. To encourage them in your garden provide them with safe nesting sites. Dense thickets of climber such as clematis, ivy and honeysuckle will encourage nesting, as will the provision of nesting boxes (with small, 28 mm, entry hole). Make sure the boxes are well hidden, giving privacy for the nervous tenants and protection from predators like domestic cats.

Hoverflies are eager flower visitors so attract them with an array of nectar-rich species and they will reward you by egg-laying in any aphid colonies. Useful hoverfly plants include: angelica, fennel, aster, yarrow, golden rod, ornamental thistles, orpine (ice plant), and there are many others. They also love buddleia, which although very vigorous can be controlled by regular hard pruning.

There are a wide choice of squirrel-proof bird feeders available and a variety of bird feeds. Peanuts are popular with many birds including the nuthatch (above) but try black sunflower seeds, fat balls or one of the many specialist mixes available to attract a variety of birds.

Traditionally, marigolds were planted in vegetable patches to 'drive off' pests. They may not have quite this effect, but they do work by attracting hoverflies to the flowers and ladybirds to the small aphid colonies. These then move to attack any aphids on nearby crops.

Other aphid predators can also be attracted by offering them early-season greenfly food on native plants such as stinging nettles. Nettle aphids won't attack garden plants, but aphid predators like ladybirds, lacewings, hoverflies and ground beetles will move out into the rest of the garden as

An area for **wildlife** – the **basics**

- Wildlife needs food, water, shelter and places to breed.
- Native wildlife thrives best on native plants (including 'weeds').
- Attract wildlife variety by a varied planting of low/medium/tall climbing plants.
- Avoid chemicals: sprays, powders and pellets.
- Don't over-fertilise the soil – use only home produced and organic compost and mulches.
- Don't be too tidy *everywhere* in the garden.

A small log pile gives shelter to many predators like toads, frogs and ground beetles.

the season progresses. Leave a portion of the garden to run wild, and let nettles sprout. Cut the nettles to encourage the predators to disperse and to prevent the nettles from becoming too established.

A mulch of bark chippings or compost around plants will help reduce moisture loss and encourage predators like centipedes and ground beetles. Diseased plants should be cut and burned and not added to the compost bin.

A small log pile gives shelter to many predators like toads, frogs and ground beetles. Leave it during the winter to allow safe hibernation, but examine it regularly on summer days to clear out any slugs and snails.

habitats for wildlife

- **Borders** – Choose a mixture of seeds, bedding and established plants for immediate effect.
- **Grass** – If establishing a new lawn, seeds provide an opportunity for a more sympathetic mixture of species. Ease the mowing regime and remove grass cuttings. Alternatively, set-aside an area for wild flowers and grasses.
- **Hedges** – Use traditional hedgerow species such as hawthorn, blackthorn, hazel, holly, beech, lime, viburnum, spindle, privet and yew.
- **Log piles** – Keep unwanted wood from felled trees and lopped branches. Stack irregularly for shelter and fungus-feeders.
- **Shade and glades** – Use the variety of sunny and shady areas to encourage sun-loving or woodland glade plants.
- **Shrubs** – Plant them where they will have room to expand.
- **Trees** – Plant native species like birch, rowan and wild cherry to attract wildlife and let in light. Avoid fast-growing species such as ash, sycamore and leyland cyprus.
- **Trellis and fences** – A trellis or fence planted with climbers provides effective wildlife friendly shelter and screening.
- **Water** – A pond or even a half-barrel will attract aquatic insects to breed and birds to drink.

Put in a **pond?**

Help from a **pond**

Putting in a pond will bring some of the best help to bear on any pests you have. Frogs and toads eat slugs, dragonflies eat flying insects. Choose a sunny open spot, not overhung by trees or bushes, but with a hedge or fence nearby to offer some shelter. Using a flexible plastic or butyl (rubber) liner is the easiest means of construction, or if space is very tight a pre-formed plastic pond. For more information on how to build a pond see the *Green Essentials* title, *Create Ponds*.

Putting in a pond is one of the most effective methods of attracting wildlife allies.

Some flowers to grow to **attract pest eating insects**

Hoverflies, wasps, ladybirds and flower bugs are some of the insect predators that will help control pests - attract them by planting some of the following near your vegetable patch or vulnerable plants:

- Cornflower
- Gazania
- Corn marigold
- Sunflower
- Fennel
- Nemophila
- Californian poppy
- Poached egg plant
- Pot marigold
- Annual convolvulus
- French marigolds
- Yarrow

Companion planting

Companion planting is another organic technique for keeping pests and diseases at bay. The theory is that certain combinations of plants can help growth and reduce pest or disease problems. Companion planting works in different ways, for example:

- repelling insects by giving off natural chemicals or scent
- attracting beneficial insect predators and pollinators
- as decoys to keep pests away from a vulnerable main crop
- absorbing minerals and/or fixing nitrogen in the soil
- creating shade or a windbreak and providing ground cover.

Companion planting is worth trying, with the proviso that there is no hard and fast evidence that all the 'companionships' suggested in books are effective. Try out some of the suggestions below making sure the companions don't compete for food and water., but do be prepared to use other organic methods if needs be.

French marigolds to repel whitefly from greenhouse tomato crops.
Onions to repel carrot fly.
Sweetcorn to provide shade for summer lettuce.
Interplanting cabbages with an unrelated plant such as french beans to repel aphids and whitefly.
Dill to attract hoverflies and wasps which eat aphids.
Garlic as deterrent for aphids, this is particularly good to plant with roses.

action stations

1. **Be vigilant.** Check your garden regularly looking for chewed leaves, blotched foliage or wilting stems.

2. **Try to avoid planting plants that you know are vulnerable to attack.** If you have too much trouble with a particular pest on a specific plant, consider giving up gracefully and growing something else, rather than resorting to the chemical pesticides.

3. **Protect new plants.** Give new plants a good start by planting as per the instructions into well prepared soil. If necessary, protect with barriers such as netting and cloches.

4. **Encourage pest predators.** Create some habitat in which predators can live and breed. Add nesting boxes for birds and consider putting in a pond which will attract many predators including frogs, toads and dragonflies.

green essentials – **organic guides**

- **Attract Wildlife**
- **Create Compost**
- **Grow Fruit**
- **Perfect Lawns**
- **Banish Slugs**
- **Create Ponds**
- **Grow Vegetables**
- **Perfect Roses**
- **Control Pests**
- **Garden Birds**
- **Healthy Plants**
- **Successful Allotments**

For more information please visit

www.impactpublishing.co.uk

Impact Publishing Ltd 12 Pierrepont Street Bath BA1 1LA • Tel: 01225 446666

who, what, where, when and why organic?

for all the answers and tempting offers go to www.whyorganic.org

- Mouthwatering offers on organic produce
- Organic places to shop and stay across the UK
- Seasonal recipes from celebrity chefs
- Expert advice on your food and health
- Soil Association food club – join for just £1 a month

Resources

HDRA the organic organisation promoting organic gardening farming and food
www.hdra.org.uk
024 7630 3517

Soil Association the heart of organic food and farming
www.soilassociation.org
0117 929 0661

Pesticide Action Network UK
www.pan-uk.org

MAIL ORDER:

The Organic Gardening Catalogue Organic seeds, composts, raised beds, barriers, traps and other organic gardening sundries. All purchases help to fund the HDRA's charity work.
www.organiccatalogue.com
0845 1301304

Agralan Ltd Traps, copper tape
www.agralan.co.uk 01285 860015

Centre for Alternative Technology Various natural pest control devices
www.cat.org.uk 01654 705950

Garland Products Ltd Traps
www.garlandproducts.com
01384 278256

Green Gardener Various slug and snail repellent devices
www.greengardener.co.uk
01394 420087

Tamar Organics Various natural pest control devices also organic seeds
www.tamarorganics.co.uk
01822 834887

Want more organic gardening help?

Then join HDRA, the national charity
for organic gardening, farming and food.

As a member of HDRA you'll gain-

- free access to our Gardening Advisory Service
- access to our three gardens in Warwickshire, Kent and Essex and to 10 more gardens around the UK
- opportunities to attend courses and talks or visit other gardens on Organic Gardens Open Weekends
- discounts when ordering from the Organic Gardening Catalogue
- discounted membership of the Heritage Seed Library
- quarterly magazines full of useful information

You'll also be supporting-

- the conservation of heritage seeds
- an overseas organic advisory service to help small-scale farmers in the tropics
- Duchy Originals HDRA Organic Gardens for Schools
- HDRA Organic Food For All campaign
- research into organic agriculture

To join HDRA ring: 024 7630 3517
email: enquiries@hdra.org.uk
or visit our website: www.hdra.org.uk

Charity No. 298104